Environmental Awareness: Water POLLUTION

AUTHOR
Mary Ellen Snodgrass

EDITED BY
Jody James, Editorial Consultant
Janet Wolanin, Environmental Consultant

DESIGNED AND ILLUSTRATED BY
Vista III Design, Inc.

BANCROFT-SAGE PUBLISHING, INC.
601 Elkcam Circle, Suite C-7, P.O. Box 355
Marco, Florida 33969-0355

Library of Congress Cataloging-in-Publication Data

Snodgrass, Mary Ellen.
Environmental awareness—water pollution / by Mary Ellen Snodgrass; edited by Jody James, Editorial Consultant; Janet Wolanin, Environmental Science Consultant; illustrated by Vista III Design.
p. cm.—(Environmental awareness)
Includes index.
Summary: Discusses the importance of a clean water supply and how water pollution threatens the lives and health of people, animals, and plants. Readers learn how to contribute to anti-pollution and conservation efforts.
ISBN 0-944280-26-9
1. Water—Pollution—Juvenile literature. 2. Water conservation—Juvenile literature. [1. Water—Pollution. 2. Water conservation. 3. Conservation of natural resources.] I. James, Jody, Wolanin, Janet. II. Vista III Design. III. Title. IV. Title: Water pollution. V. Series: Snodgrass, Mary Ellen. Environmental awareness.
TD422.S62 1991
363.73'94—dc20

International Standard Book Number:
Library Binding 0-944280-26-9

Library of Congress Catalog Card Number:
90-20949
CIP
AC

PHOTO CREDITS

COVER: Vista III Design; The Bettmann Archive p. 29; Nancy Ferguson p. 20; Firth Photobank, Bob Firth p. 4; Robert Koontz p. 43; K.G. Melde p. 21; Minnesota PCA p. 36; Silver Image, Steve Hobbs p. 7; Unicorn Photography, D&I McDonald p. 16; USAF p. 39; Vista III Design, Ginger Gilderhus p. 23, 30, Grant Gilderhus p. 8, 11, 12, 19, 22, 25, 26, 27, 28, 33, Jackie Larson p. 40; Wide World Photos, Inc. p. 35.

TABLE OF CONTENTS

Ozark Mountains

CHAPTER ONE

A THREAT TO CLEAN WATER

The health of the earth depends on a supply of clean water. Without water, people could not live. Animals and plants would also die. As good citizens, we must try to keep our water clean. Unfortunately, many of us are not doing a good job. Careless people in factories and homes cause a great deal of water **pollution.** Because of their carelessness, rivers and streams often become too dirty for recreation.

Here is what happened to the Bartrams, who drove to the Ozark Mountains for their summer vacation.

AN UNPLEASANT SURPRISE

The afternoon sun was setting behind the Ozarks. The Bartrams' camper nosed its way through heavy traffic. In the back seat, Jill studied the map to see how far they must drive before they arrived at their favorite vacation spot. It was her job to help her parents watch for signs and directions.

"We're almost there, Mom. Only one more turn," she said with excitement in her voice. "Look for the signpost."

Jill's parents watched the road ahead for the sign saying "Raider's Point." In a few minutes, they could see the tops of tents and campers to the left of the main road. Beyond, they could hear the gurgle and splash of running water.

"Look, Mom," Jill said, "there's Raider's Creek. We can almost fish from our tent door."

Mrs. Bartram laughed. "We're on vacation, Jill, but I hope we're not that lazy!"

"Don't speak too soon," Jill's father smiled. "I'm feeling slow and lazy. Maybe Jill has the right idea."

The Bartrams parked in the shade of two big pine trees. Mr. and Mrs. Bartram began unloading baskets of food and putting up the tent. Jill helped by removing the fishing gear from the camper.

"Let's hurry," Jill yelled excitedly. "I want fresh rainbow trout for dinner."

After their tent was set up, the family changed into jeans and flannel shirts. Then the Bartrams got their fishing rods. Mr. Bartram stuck a few feathered lures on the side of his hat and reached for the landing net. Together, the Bartrams wandered down the slope toward Raider's Creek.

"Gosh," Jill noted with disgust, pointing toward the trail. "Look how many tin cans, paper plates, and plastic bottles people have left behind. It wasn't this messy last year."

"We'll just have to gather some of this trash and put it in trash cans before the return trip," Mr. Bartram replied. "We want to keep coming here for a long time."

Jill dashed ahead. When she got to the edge of the stream, she stopped and gazed down at the stream. "Dad, Mom!" she called. "Come quick! What does this sign mean?"

Mr. Bartram pushed back his hat as he read the sign twice — even the small print. *"EXTREME DANGER. Stream polluted by mercury and lead. All wading, swimming, and fishing are forbidden by order of the* ***Environmental Protection Agency.***"

Jill's voice filled with sadness as she leaned her fishing rod against a pine tree. "Oh, no! We've driven more than two hundred miles to get away from the dirt and noise of the city. We've looked forward to this vacation since December. It's not fair that some people ruined Raider's Creek for the fish — and us."

Trash causes pollution in our lakes and streams.

Our earth needs water to survive. We must protect the earth's water supply from pollution.

CHAPTER TWO

WATER AND YOU

Do you agree with Jill? Do you think clean water is important? Why do factory owners allow mercury and lead to get into the water? How can the Bartrams help clean up the mess? What can you do to protect yourself from polluted water? These and other questions will help you focus your attention on the importance of clean water.

WATER FOR SURVIVAL

The earth is a special planet in our solar system. It contains a large amount of water. Three-fourths of the earth's surface is made up of water. Oceans, seas, rivers, lakes, ponds, wells, springs, streams, marshes, swamps, glaciers, and icecaps are filled with water, both fresh and salt. When this water **evaporates**, the air becomes wet, blanketing and cooling the earth below.

The earth needs water to exist. People depend on water for life. Other living things also depend on water. It helps trees and green plants to make food. Water gives fish, frogs, toads, and many types of plants an **environment** in which to live.

Without water, this would be a very different world. It would be a place filled with rocks, deserts, mountains, and little else. The sun would rise and set. The moon would continue to shine. But no flowers, trees, or grass would grow. The winds would blow, but no birds would sing.

If water became scarce or unusable, it would cause terrible, deadly situations. Farmers would lose valuable crops. People would go hungry and thirsty. Animals would be unable to find grass to eat or streams from which to drink.

WATER AND THE HUMAN BODY

Water is very important in keeping the body strong. It helps soften and strengthen skin and organs such as the heart and lungs. Water assists in **digestion** and **respiration**. It helps carry food to our limbs and oxygen to our bloodstreams. It also gives pleasure and exercise to swimmers and boaters.

Without food and drinking water, human bodies would begin to shrink. Muscles would become weak. People would be unable to work or play. If people go without proper amounts of food and water for a long time, their minds and bodies can be damaged. This is especially true for small babies and young children.

Dehydration, or loss of water, is very serious. Water forms a major part of living cells. Without it, our internal organs are unable to use food properly. Blood contains a large amount of water, too. Without water, blood cells could not travel through the body. The nerves and brain would stop working.

Water is the key to the survival of living things. Within hours, particularly during hot, dry weather, lack of water can cause unconsciousness and even death. In order to live, humans need to protect the earth's water supply.

WATER AND THE HUMAN BODY

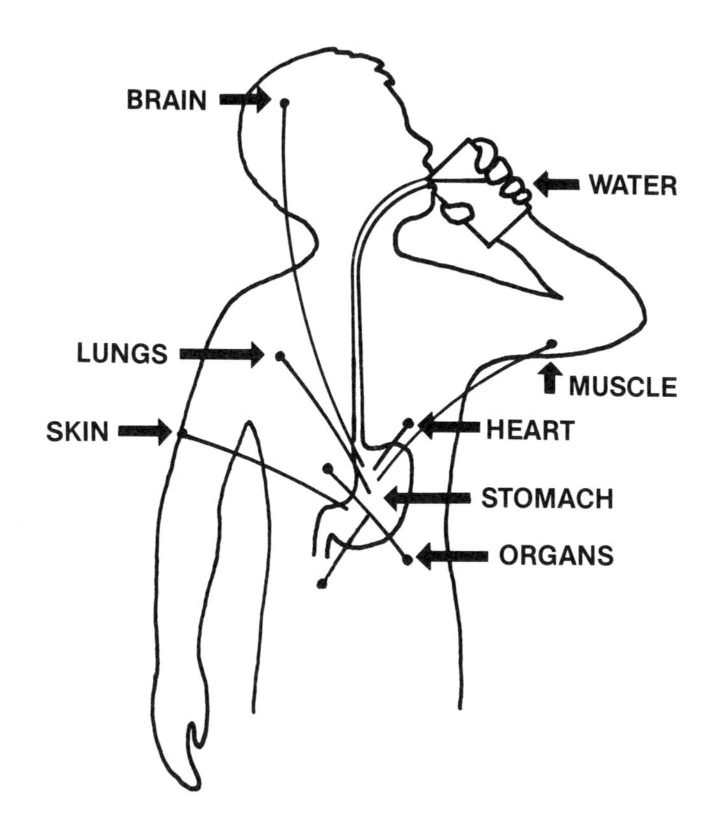

Within hours, during hot, dry weather, lack of water can cause unconsciousness and even death in people.

People who live in desert areas do not take fresh water for granted. They use only the amount of water that they need.

CHAPTER THREE

MAKING THE BEST USE OF OUR WATER

The earth's water supply is not evenly distributed. In some places, there is more than enough. In others, there is barely enough for plants, animals, and people to live.

LIVING WITH DESERTS

There are very dry regions on our earth called deserts. In fact, approximately 18% of the earth's land area is covered by deserts. In some places, such as the Sahara Desert, water is very scarce. Only specially adapted plants and animals can live there. Cacti and other **succulent** plants store the water they need in their stems and spines. Lizards, camels, and tortoises survive by storing water in their bodies.

To stay alive during the day in these dry regions, desert animals lie quietly in the shade. Some animals absorb moisture from the air. Other animals dig into the ground. They use the water stored inside their bodies. Without nature's special equipment, desert plants and animals would dry up and die.

In small towns and villages in desert areas, people do not take fresh water for granted. Families guard water supplies by digging deep wells. They cover the wells with lids or grates to keep out dirt. To stop rapid evaporation, they plant cool, protective greenery around the wells. Desert people have laws and severe penalties to stop carelessness or waste of precious water supplies.

Every day, desert dwellers use only the amount of water that they need. They use water for two or more purposes whenever possible. For example, they may use water for bathing, then reuse it to moisten plants.

MAKING SALT WATER USABLE

People in lands that are surrounded by salt water must also have fresh water to live. Salt water is harmful because it dries out the body. If people used salt water instead of fresh water for cooking and drinking, they would soon dry out.

People who live near salt water use various means of getting fresh water. For example, people in Bermuda and other Caribbean islands channel rainwater through grooves in the roofs of buildings into underground tanks. At the entrance to these tanks, screens filter out leaves and other trash. The people may add chlorine tablets to make sure the water is pure.

For nations near the sea, such as Israel and the islands of Tahiti, governments have built **desalinization** factories. These freshwater factories heat salty water. The water then forms steam, but the salt stays behind. The clean, moist steam is then separated from the salty water. The steam cools into a drinkable liquid and is drained off and pumped into public **reservoirs**.

Other types of desalinization plants use special filters, **solar energy**, or magnets to remove the salt. Unfortunately, cleaning ocean water is very expensive. Many communities cannot afford to build these factories.

EXTRACTING WATER FROM NATURE

Scientists are experimenting with methods of distributing the earth's supply of fresh water more evenly. One method is **seeding** clouds in very dry areas with a chemical called *silver iodide*. This method produces rain artificially, but it is a very expensive process. Also, rain produced in this manner is difficult to predict and control. Sometimes the rain that is made does not fall where it is needed. Other areas get too much rain.

Another way of managing weather is to slow down hurricanes and tornadoes. This method of changing nature, however, is not fully developed. Also, humans cannot predict what such changes will do to the rest of nature.

A third method that scientists have proposed is taking fresh water from glaciers and polar icecaps. Plans for use of this water are still in their early stages.

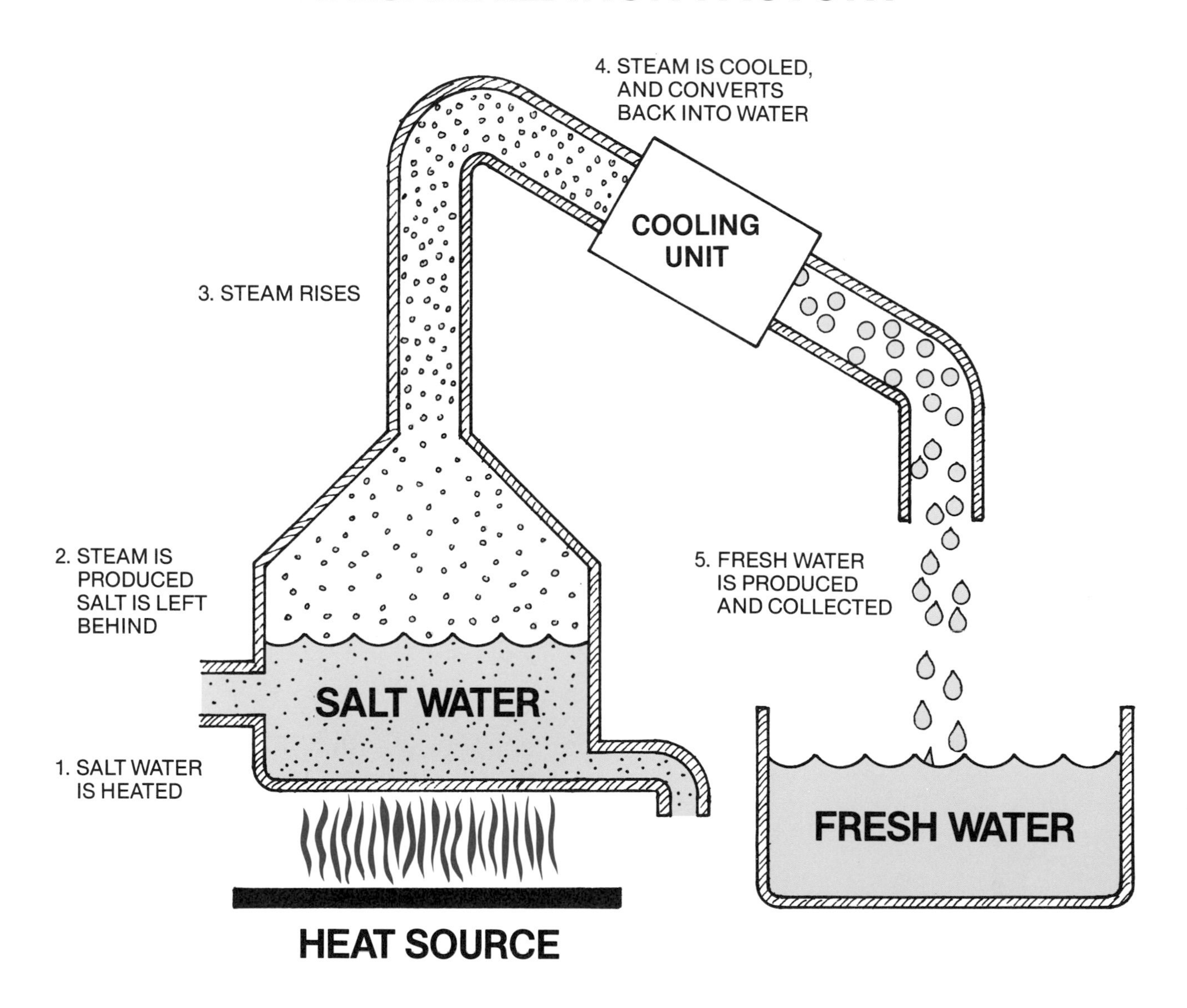
DESALINIZATION FACTORY
4. STEAM IS COOLED, AND CONVERTS BACK INTO WATER
COOLING UNIT
3. STEAM RISES
2. STEAM IS PRODUCED SALT IS LEFT BEHIND
5. FRESH WATER IS PRODUCED AND COLLECTED
SALT WATER
1. SALT WATER IS HEATED
FRESH WATER
HEAT SOURCE

When a natural disaster happens, the first job of emergency workers is to restore clean, fresh water for the people.

CHAPTER FOUR

POLLUTION AND PROGRESS

Many things threaten our world's water supply. Some of the threats come from nature. Most come from humans and their progress.

THREATS TO OUR WATER SUPPLY

Extremes of weather, such as drought and heat waves, can empty natural water basins. Violent storms and other **natural disasters** can make our water so dirty that it is not fit for human or animal use. Airborne pollutants such as dust, soot, and ash fall on water supplies. These materials can pollute city water. An overflow of salt water from oceans or spills of **fossil fuels**, such as gasoline, diesel fuel, and kerosene, can pollute local reservoirs and **aquifers**.

During natural disasters, other pollutants may also threaten our water supply. Industrial wastes, garbage, decaying plants, and the wreckage of buildings and cars clog sources of water, such as rivers. These pollutants prevent cities from getting enough drinkable water for their people.

When a natural disaster happens, the first job of emergency workers is to restore clean water. To do this, workers often truck in fresh water from distant places. When the water trucks arrive, workers measure out a supply to each person. This water is carefully sealed in clean containers. Once water is available, the workers can get on with other tasks, such as supplying food, medicine, and shelters.

HUMAN POLLUTION

The greatest threat to the earth's water supply is people. Why do people threaten their lives and the survival of the planet by dirtying their water supply?

A major cause of water pollution is industry. At one time, factories and mills were built along streams so that rapid-moving currents could turn gears to operate machinery. Factory owners also used water as a method of transportation. They floated logs downstream or loaded their products on barges. These early industries did not pollute water.

Today, however, energy comes from a variety of other sources—gas, oil, coal, electricity, steam, solar energy, and **nuclear energy**. Many of these sources of energy do pollute water.

Modern industry also has a serious problem with waste removal. Some large factories, paper mills, and oil refineries rid themselves of wastes by dumping them into streams and rivers. The pollution may move away from the factory, but it reappears downstream at another location.

Cities and businesses often haul their garbage and other solid wastes out to sea and dump them. However, the sea is filled with living animals and plants. This type of water pollution affects these animals and plants as well as the people who live along the shore.

Industrial wastes cause many changes in the environment. Nesting grounds for water animals and migratory birds, such as manatees and Canadian geese, may be overheated by hot factory wastes that are dumped into rivers and streams. Waters become too acidic or cloudy for fish to raise their young. When oysters and mussels take in pollutants, they become unhealthy for people to eat. Parks and beaches develop ugly and unhealthful heaps of trash.

Industry is not the only polluter of water resources. Individual people endanger themselves by polluting the water around them. Through careless and deliberate acts, people throw more and more garbage and other waste into every ocean, river, and stream on the planet. Wherever people live, chimneys and charcoal burners pour out smoke. Their cars use up motor oil and batteries, which are piled in **dumps**. Oil and battery acid, along with paint, insect spray, and other poisons, ooze out of these garbage dumps into fresh water.

Some large factories rid themselves of wastes by dumping them into streams and rivers.

Fly ash, which is mostly soot and dust, escapes through smokestacks where it mixes with rain and falls back to earth. There it pollutes waterways and ground water.

Another source of water pollution is from particles in the air, such as the **fly ash** produced by both homes and factories. Fly ash is mostly soot, dust, and other heavy particles that result from burning. As the fly ash passes into the air through smokestacks, it mixes with rain and falls back to the earth. There it pollutes the waterways as it is carried by rain back to streams, rivers, and **ground water**.

Even the North Pole and South Pole are not free from floating trash, such as foam products and plastics. Jacques Cousteau reported seeing ocean waste as faraway as Antarctica. All of these pollutants affect the water we drink. They also dirty the seas in which fish live.

Some pollution caused by humans results from natural **runoff**. Homeowners for example use chemicals to kill insects and to fertilize their lawns. Farmers plow hillsides. They, too, fertilize and spray crops with strong chemicals. Later, heavy rains carry loosened topsoil, **pesticides**, and chemical fertilizer downhill and into rivers, lakes, and streams.

In the same way, **strip mining** operations create heaps of unusable soil. Heavy rains carry the unprotected soil into streams and rivers.

Strip mining operations create heaps of unusable soil. Heavy rains carry the unprotected soil into streams and rivers.

Road salting and snow removal in cities in the northern part of our country pollute the water supply.

Strip mines also produce **by-products**, which are the leftover metals of their operation. These wastes may end up polluting many water sources, including faraway ground water and wells.

Cities cause water contamination in a variety of ways. Water can be polluted by fire fighting, road salting, snow removal, street sprinkling, irrigation, and leaks in water and sewer mains. Cities build water treatment and sewage treatment plants to help control these pollutants. Sometimes, however, these plants are unable to do all the work needed to clean the water. They cannot keep up with rising populations, wars, accidents, or natural disasters, especially volcanoes and earthquakes.

When treatment plants fail, the result is often terrible. Human wastes, carrying foul tastes, odors and deadly bacteria, overrun sewer plants and pour into streams and oceans. These filthy waters dirty backyards, parks, and beaches. People can suffer major and minor illnesses from polluted waters, such as food poisoning, dysentery, typhoid, cholera, and meningitis.

Cities build water and sewage treatment plants to help control pollutants caused by firefighting, snow removal and leaks in water and sewer mains.

PROGRESS AND THE WATER SUPPLY

Human progress is a threat to water. When asphalt and concrete roads and parking lots replace trees, nature is thrown off balance. Water can no longer soak into the soil.

As more lands are cleared for homes and businesses, waterways suffer immediate danger of permanent damage. The **wetlands** of Florida and North Carolina, for example, lose water as people cut trees and brush. Homes replace open swamps. An increase in traffic and industry produces a danger to the water supply. More pollutants from exhaust fumes, sewage, and smoke seep into the soil. Even with good water treatment facilities, many citizens must try to protect themselves further from water pollution.

The development of newer and tougher types of plastics and other manufactured goods leads to types of garbage that do not break down or decay in landfills. These wastes include disposable diapers, packaging materials, and storage bags. Many artificial goods resist methods of disposal and waste management, such as burial or **incineration**. As a result, runoff carries more of these pollutants into the water supply. Other pollutants are deliberately dumped into waterways.

More serious than home products are hospital disposables and **toxic wastes**. These wastes come from hospitals, fuel dumps, military bases, nuclear power plants, and chemical refineries. Like some home wastes, they do not break down in the soil. Runoff from these wastes can carry serious diseases, such as hepatitis or staph infections. It can also spoil ground water with mercury, lead, and other heavy metals.

As more land is cleared for homes and businesses waterways suffer damage. For example, the wetlands of Florida lose water as people cut trees and brush.

As the population of our earth increases, the amount of waste created by these people increases. This causes more pollution and disposal problems.

CHAPTER FIVE

KEEPING OUR WATER SUPPLY CLEAN

As the planet grows more crowded, people and animals produce more waste. This problem grows worse in wealthy areas, such as Japan, Europe, and the United States. Because people in these areas use more products, they have more waste to dispose of.

Also, because people live closer together, their **hygiene** problems increase. There must be careful control of waste so that the supply of water will be kept pure. If the water supply becomes polluted, disease can spread rapidly.

However, cleanup costs are high. How can cities remain livable and people stay healthy without spending too much money? Large groups of city planners and scientists work daily on the answer.

As people are forced to live closer together, hygiene problems increase.

CLEAN WATER IN PAST CENTURIES

History gives some idea of how ancient peoples solved water pollution problems. For centuries, large cities, such as ancient Rome and medieval London, protected their water supplies with simple measures. City planners built community wells far from sewage. Even though they did not understand the connection between germs and foul waters, they drained swamps to halt diseases spread by insects.

To supply citizens with water, the city planners turned to cleaner sources. From melted snow in the mountains, they channeled fresh water downhill through **aqueducts**. In these areas far away from the city there were fewer human and animal pollutants. Thus, the water was cleaner. To help keep water clean, city workers picked rivers and springs clean of trash, garbage, and decaying animals and fish.

During the nineteenth century, Louis Pasteur discovered how germs spoil water supplies. At his suggestion, people began to boil drinking water they got from community **hydrants** and wells. They stored the clean water in scalded containers. These methods helped control outbreaks of plague, tuberculosis, dysentery, cholera, and typhoid.

To supply citizens with water, melted snow was channeled from the mountains downhill through large aquaducts.

For centuries, cities such as London, England built community water wells far away from sewage pollution. This helped stop diseases.

By using contour plowing, farmers can keep more of the chemical fertilizers and insecticides on the hillsides and out of the creeks and streams.

MODERN POLLUTION-FIGHTING MEASURES

In the early years of the twentieth century, city leaders improved the public water supply by protecting **watersheds**. Watersheds are natural pathways through which rainwater travels on its way to streams and reservoirs.

Farmers began using **contour plowing**, or plowing by following the curve of a hill sideways rather than up and down the slope. In this way, farmers could keep more of the chemical fertilizer and insecticide on the hillside and out of creeks and streams.

Loggers became more careful about cutting timber. They left a greater amount of brush behind to cover the ground. By leaving growing plants on the forest floor, they helped slow the flow of water. More of the slow-moving water then soaked into the soil. By these methods, people helped to reduce the amount of runoff that polluted the watersheds. Therefore, towns and cities had a cleaner water supply.

People also began to save ground cover and keep watersheds clean in other ways. Some ways include the prevention of forest fires, **soil erosion**, overgrazing by cattle and sheep, and other ruin of ground cover. By improving ground cover, people can save water's natural filter and help stop pollutants from entering the watersheds.

Overgrazing by sheep and cattle must be carefully controlled. The animals can destroy ground cover, allowing pollutants to enter the water supply.

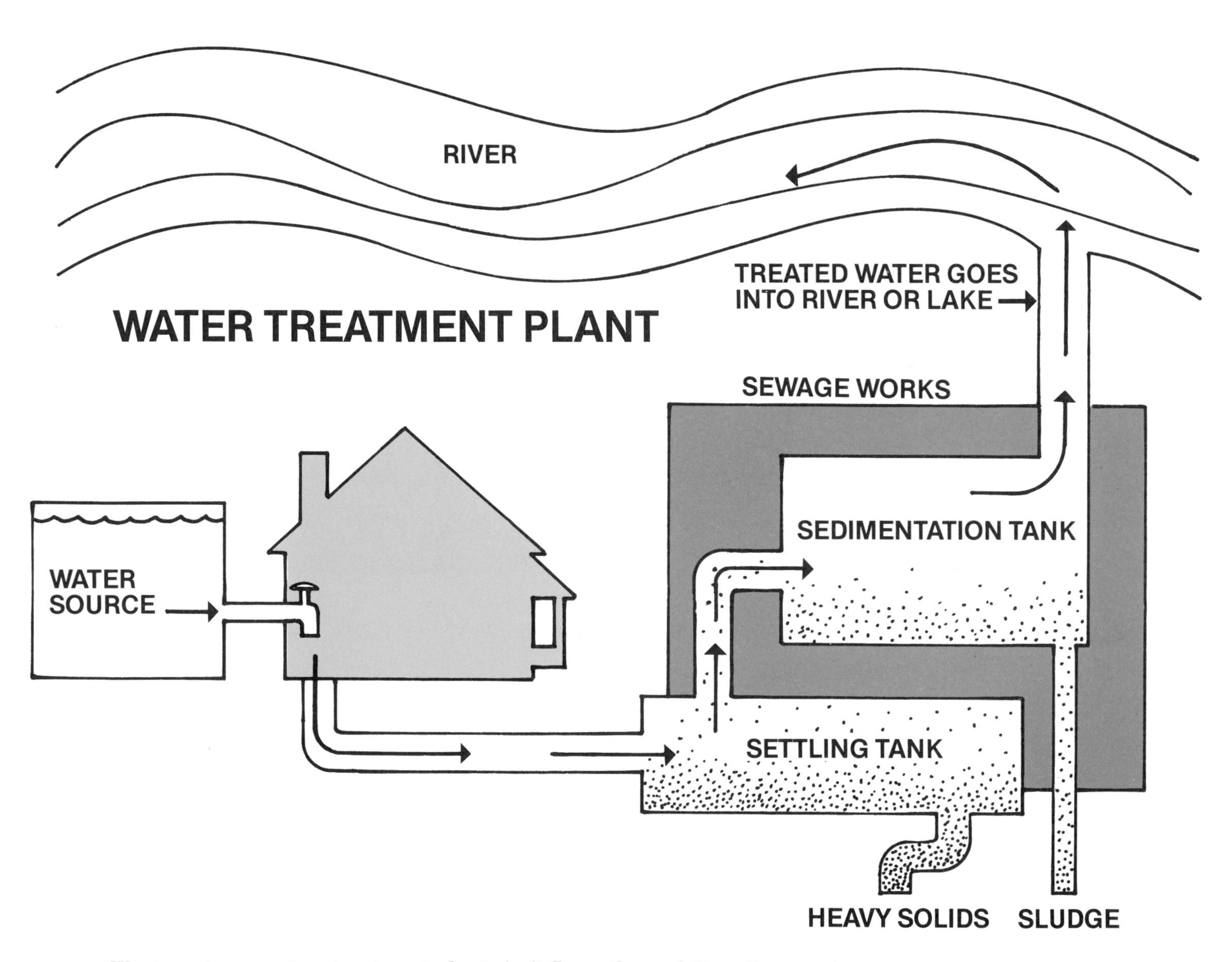

Waste water goes to a treatment plant. As it flows through it, sedimentation tanks filled with sand or gravel trap pollutants.

CLEAN WATER TODAY

Today, water is treated in several ways to make it clean. One method is to pipe it through filters and **sedimentation** tanks filled with gravel or sand. The closely-packed grains of sand or small rocks trap pollutants like a strainer as they flow through.

Another method of water treatment is **aeration**, in which air bubbles through water or water sprays into the air. This method improves the taste of water and removes stain-producing metals, such as iron.

A third method is the use of **ozone gas**. Water treatment workers force this powerful gas through drinking water to remove germs. Ozone gas also improves both the taste and smell of water.

Water treatment workers also use other chemicals to clean water. They add ammonia, fluoride, and chlorine to control slime, which causes bad odors and tastes in water.

When outbreaks of disease suggest that water is unclean, **sanitation engineers** take samples of the water. These public health agents then treat the problem. They sample water daily until all danger passes. Their careful watch also provides valuable facts about clean water and its relation to health.

Another part of water treatment is **water softening**. This method removes minerals that make water "hard." Hard water rusts and destroys pipes. These leaky pipes allow pollutants to seep into underground water supplies.

Hard water rusts and destroys pipes. These leaky pipes allow pollutants to enter the underground water supply.

PROTECTING WATER SUPPLIES

Protection of water supplies is a never-ending task. The Environmental Protection Agency (EPA) is a government office that guards our country's water and other natural resources. This agency keeps a close eye on factories that often pollute the water supply. The EPA is especially watchful of dangerous runoff that can pollute our rivers and streams. Large ships that carry oil products and dangerous chemicals, such as explosives, are often inspected by the EPA for leaks and spills.

This agency tries to prevent accidents caused by companies that drill for oil beneath the ocean floor. Because oil spills threaten beaches, sea animals, plants, and fish, volunteer groups remain alert to these spills. They keep supplies at hand to trap oil before it can reach the shore.

The ocean also will slowly help clean itself by churning large clumps of spilled petroleum with sand and other particles. These clumps will soon sink to the bottom of the ocean where they can be gathered and removed.

There are other methods of stopping oil spills from spreading. Workers float bales of straw to keep the spill in one place. Then they use oil-eating substances in oil-coated waters. They sometimes use powdered chalk to absorb the oil and stop its spread.

When oil reaches land, many methods are tried to restore cleanliness. Using hot water from large hoses, workers squirt rocks and beaches to remove oil from them. Sadly, all of these methods are unable to remove all the oil from the beaches.

Large ships carrying oil products and dangerous chemicals are inspected by the Environmental Protection Agency for leaks and spills.

A thick layer of crude oil coats the feathers of a Cormorant in Northern Saudi Arabia. Thousands of barrels of oil were dumped into the Persian Gulf from Kuwati refineries.

Water tests help to discover heavy metal particles. If allowed to dirty our water, these metals would poison our body tissues.

CONTROLLING AIRBORNE WATER POLLUTANTS

There are several methods of stopping water pollution from airborne particles, such as fly ash. Regular air, soil, and water tests help to discover the spread of dangerous **heavy metal** particles, including lead, zinc, chromium, arsenic, and mercury. If allowed to dirty our water, these metals would poison our body tissues.

To control dangerous particles, government rules now make factory owners build taller smokestacks. These tall pipes stop pollutants from falling directly to the ground. However, this method does not remove the ash. It only spreads the ash over a greater area.

Factories have other ways of halting fly ash. Some factories place covers over the smokestack to trap particles. Others mix steam with the fly ash to make it fall back down the smokestack. In this way, the escaping gas is cleaned and particles are kept from filling the air and dirtying water.

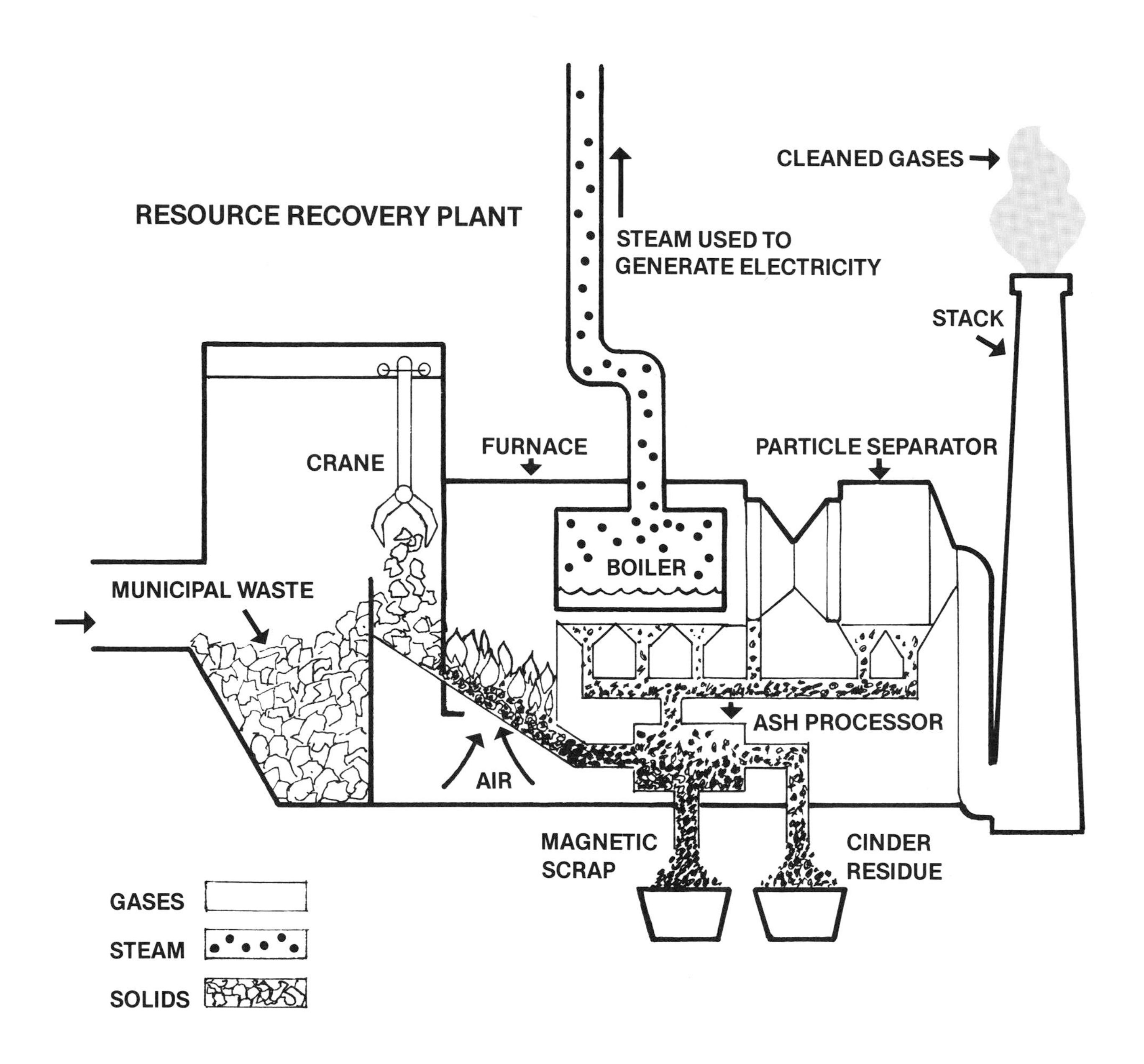

This plant mixes steam with fly ash to keep the escaping gases clean.

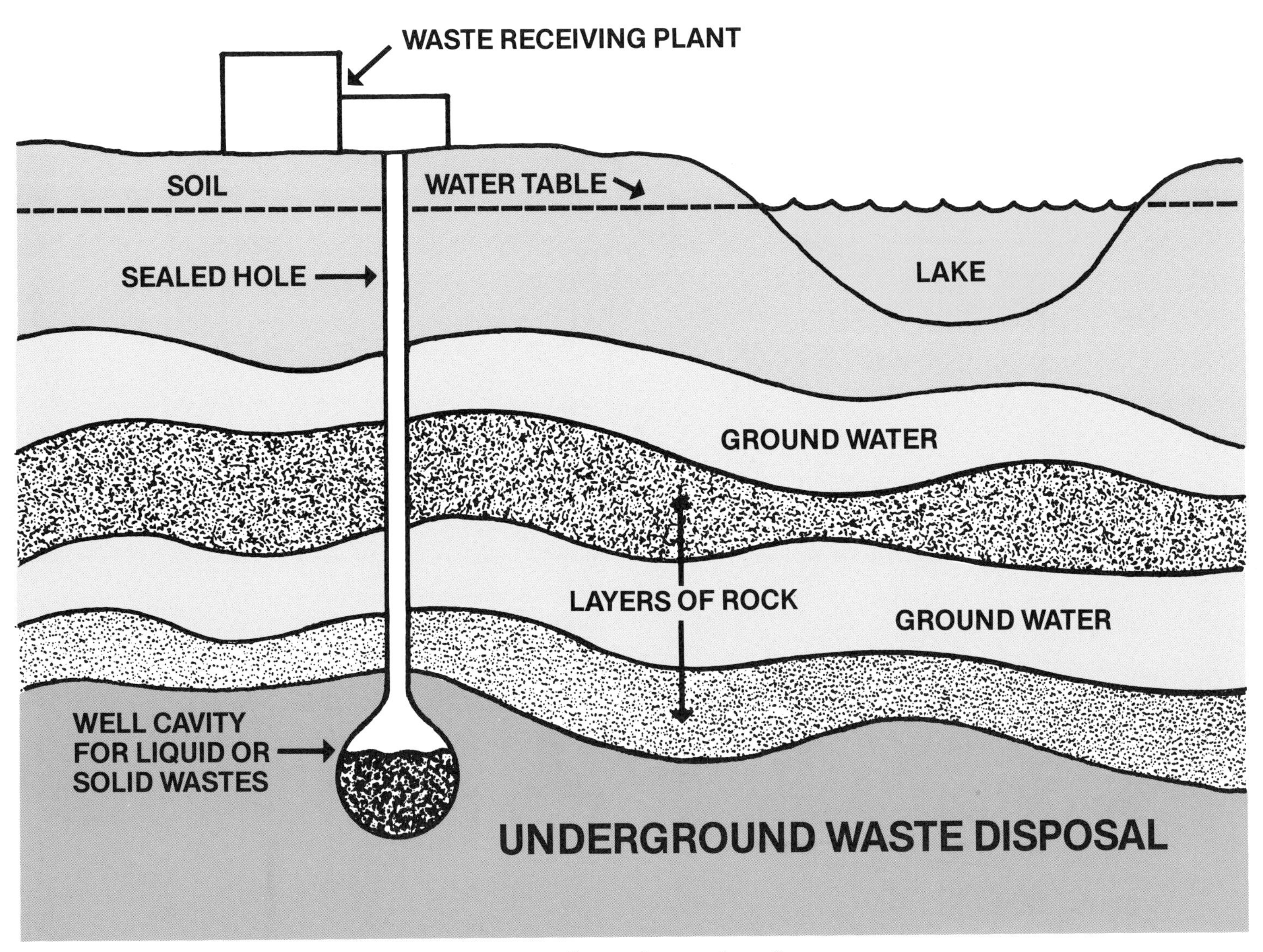

Wastes injected into deep wells cannot return to the surface and spoil ground water or the water table.

CREATIVE METHODS OF DISPOSAL

Government agencies work hard to find creative, safe, and inexpensive ways to dispose of waste. One solution is through the sale of recycled paper, cardboard, aluminum, and glass. If all citizens join in recycling these useful materials, cities can sell them to factories where they are reused. The money the cities earn can be used to improve water treatment.

Another way to dispose of wastes is to inject them into deep wells. Drillers dispose of pollutants in holes 1,000-12,000 feet deep. Trapped under layers of rock, these pollutants cannot return to the surface and spoil ground water or the **water table**. This method is cheap and convenient. It also ends the problem of moving waste through our neighborhoods by truck or rail. This method, however, is not perfect. Some sources suggest that an earthquake might release the pollutants into the ground water.

Another proposed method is to dump waste into outer space, far beyond our air supply. By loading pollutants on rockets, scientists can shoot the waste far away from water supplies. However, not everyone agrees that this method is a good idea. Many pollution experts warn that people on earth must not use outer space as a dump. Such a method only moves the problem from one place to another. This method of disposal is also expensive and can only dispose of small amounts of waste at a time.

A proposed method of disposal is to dump waste into outer space. Not everyone agrees that this method is a good idea as it only moves the problem from one place to another.

Do not water lawns and gardens during dry periods.

CHAPTER SIX

THE INDIVIDUAL'S PART

One person cannot halt pollution. Still, each person can be more responsible for stopping the waste or pollution of precious water supplies. Think back to the experience the Bartrams had in the Ozarks. To protect our campgrounds and parks, as well as our drinking water supply, everyone must help. Here is a list of suggestions that anyone can use:

SAVE WATER

1. Turn off faucets when water is not in use, such as while brushing teeth or washing hair.
2. Place a container filled with sand in the toilet tank to decrease the amount of water used for flushing.
3. Do not water gardens and lawns or wash vehicles during dry periods.
4. Recycle bathwater, dishwater, and washing machine flow. Use it to water houseplants and garden plots and to rinse driveways, porches, and patios.
5. Shorten showers or place a restrictor on the shower head to reduce water use.

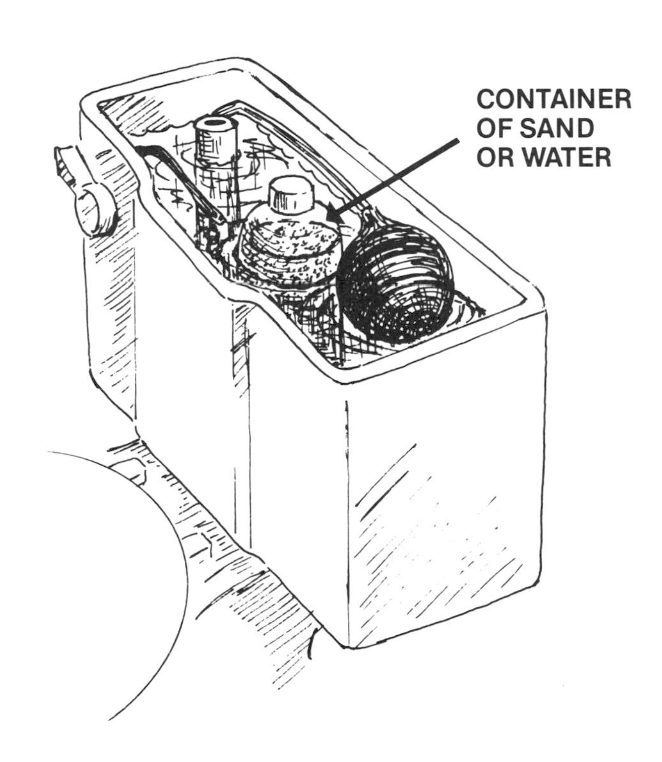

PROTECT WATERWAYS

1. Remove trash from roadways, recreation areas, stream banks, and beaches.
2. Make certain that paper goods and other debris do not blow away from boats, picnic tables, and cars.
3. Establish containers for recycling aluminum cans, glass, plastic, paper, and cardboard. Make frequent trips to a **recycling center** so that these goods can be reused rather than dumped where they may pollute ground water.
4. Dispose of paint, used oil, pesticides, and other pollutants according to local laws.
5. Avoid the use of products that contain large amounts of phosphorous. This chemical helps choke waterways by encouraging the growth of water plants.
6. Insist on **biodegradable** products which do not pollute water supplies.

PROTECT YOURSELF

1. Swim and fish only where water is clean. Drink and bathe only in tested water from wells, springs, and city pipelines. Watch for signs warning of pollution.
2. Never eat fish or shellfish caught in polluted water.
3. If you have a question about the taste, smell, or appearance of a water supply, call your local health department for a free test. You can also write to the state department of natural resources for more information.
4. On camping trips, boil water for drinking and cooking, use **decontamination** tablets to kill bacteria, or buy a special pump to filter out pollutants.
5. Visit water and sewage treatment plants to learn how cities protect citizens from water pollution.

On camping trips, boil water for drinking and cooking to help kill bacteria.

Read articles about pollution.

ENCOURAGE GOVERNMENT OFFICIALS TO TAKE ACTION

1. Write or call local officials or state or national representatives. Encourage them to vote for stronger anti-pollution laws. Insist that they enforce the laws we already have. Insist on heavy fines for polluters.
2. Support citizen's groups, such as those that clean stream banks and beaches.
3. Observe streams, lakes, rivers, and ponds near your home. Report unusual changes, such as fish kills, dumping, floating trash, oil slicks, or foul smells.

ENCOURAGE INDUSTRY TO KEEP WATERWAYS CLEAN

1. Avoid products made by manufacturers that pollute or refuse to cooperate in the cleanup and protection of waterways.
2. Report dumping and dangerous or suspicious runoff near streams and sources of drinking water.
3. Ask questions. Read articles about pollution. By learning more, you can be an effective voice in cleaning our water supply.

GLOSSARY

aeration (ay RAY shuhn) adding air to a substance

aqueducts (AK wih duhktz) an elevated pipe that carries water a long distance

aquifers (AK wih fuhrs) a layer of underground rock or gravel that stores water

biodegradable (by oh dih GRAYD uh buhl) capable of breaking down into harmless products

by-products (BY prahd uhkts) waste products made by a factory while it is creating useful goods

contour plowing (KAHN tuhr PLOW ihng) following the natural curve of a slope sideways across a field to prevent gullies and decrease erosion

decontamination (dee kuhn tam ih NAY shuhn) cleansing; purifying

dehydration (dee hy DRAY shuhn) the loss of body or plant fluids, especially water

desalinization (dee sal ih nih ZAY shuhn) removal of salt from water

digestion (dy JEHS chuhn) the process of breaking down food into simple substances so that the body can use them

dumps (DUHMPS) a place where garbage is piled up in the open

environment (ihn VYRN mihnt) the world in which living things live and grow

Environmental Protection Agency (EPA) a federal agency that helps protect the environment for United States citizens

evaporates (ee VA puh rayts) to change from a liquid to a gas

fly ash (FLY ASH) solid particles of dust, soot, and ash which are blown from a fire

fossil fuels (FAHS ihl FYOOLZ) fuels such as coal, oil, and natural gas that formed over millions of years from the decay of dead plants and animals

ground water (GROWND waht uhr) water that collects below the earth's surface

heavy metal (HEHV ee MEHT uhl) metals, including iron, zinc, mercury, lead, chromium, and arsenic, some of which can poison humans and animals

hydrants (HY druhnts) city water main; water tap

hygiene (HY jeen) habits that lead to good health

incineration (ihn sihn uh RAY shuhn) burning dangerous waste to reduce it to ash

natural disasters (NACH [uh] ruhl dihz AS tuhrz) severe environmental problems that have natural causes such as earthquakes, floods, violent storms, and volcanoes

nuclear energy (NOO klee uhr IHN uhr jee) energy that is stored in atoms

ozone gas (OH zohn GAS) a powerful gas that deodorizes and purifies

pesticides (PEHS tih syds) a chemical that kills insects and other pests

pollution (puh LOO shuhn) dirtying of the environment

recycling center (ree SY klihng SIHN tuhr) a place where reusable materials, such as newspaper, glass, or aluminum cans, are collected and sorted for reuse

reservoirs (REH suhrv oyrz) an above ground place where water is stored for future use

respiration (rehs puh RAY shuhn) the way that plants and animals supply cells with oxygen and carry away waste gas

runoff (RUHN ahf) soil particles that are carried downhill by rain or melting snow

sanitation engineers (san ih TAY shuhn ihn juh NEERS) a public employee who promotes good health by maintaining clean conditions and preventing disease

sedimentation (sehd ih mihn TAY shuhn) a process that causes heavy solids to sink to the bottom of a liquid

seeding (SEED ihng) putting a chemical in clouds to make it rain

soil erosion (SOYL ee ROH zhuhn) the loss of topsoil after roots and other supports are destroyed

solar energy (SOH luhr IHN uhr jee) energy derived from the sun

strip mining (STRIHP MYN ihng) a mining operation that strips away soil to uncover a valuable mineral, such as coal

succulent (SUHK yoo luhnt) a plant made up of fleshy, water-filled tissues

toxic wastes (TAHK sihk WAYSTS) wastes that can poison living things

watersheds (WAHT uhr shehds) a natural pathway for rainwater

water softening (WAHT uhr SOF nihng) a chemical process that removes heavy metals from water

water table (WAHT uhr TAY buhl) the portion of the ground which is filled with water

wetlands (WEHT landz) marshes, swamps, bogs, ponds, and lakes